MACK
FG-FH-FJ-FK-FN-FP-FT-FW
1937 through 1950
PHOTO ARCHIVE

MACK
FG-FH-FJ-FK-FN-FP-FT-FW
1937 through 1950
PHOTO ARCHIVE

Photographs from the
Mack Trucks Historical Museum Archives

Edited with introduction by
Thomas E. Warth

Iconografix
Photo Archive Series

Iconografix
PO Box 609
Osceola, Wisconsin 54020 USA

Library of Congress Card Number 95-77484

ISBN 1-882256-35-2

95 96 97 98 99 00 5 4 3 2 1

Book and cover design by Lou Gordon, Osceola, Wisconsin

Printed in the United States of America

Book trade distribution by Voyageur Press, Inc. (800) 888-9653

PREFACE

The histories of machines and mechanical gadgets are contained in the books, journals, correspondence and personal papers stored in libraries and archives throughout the world. Written in tens of languages, covering thousands of subjects, the stories are recorded in millions of words.

Words are powerful. Yet, the impact of a single image, a photograph or an illustration, often relates more than dozens of pages of text. Fortunately, many of the libraries and archives that house the words also preserve the images.

In the Photo Archive Series, Iconografix reproduces photographs and illustrations selected from public and private collections. The images are chosen to tell a story—to capture the character of their subject. Reproduced as found, they are accompanied by the captions made available by the archive.

The Iconografix Photo Archive Series is dedicated to young and old alike, the enthusiast, the collector and anyone who, like us, is fascinated by "things" mechanical.

ACKNOWLEDGMENTS

The photographs appearing in this book were made available by the Mack Trucks Historical Museum. We are grateful to Colin Chisholm, Curator, for his assistance.

Paul Volppe's rendition of a Mack FN. One of a series of six "American Heritage" commemorative pewter plates produced in the 1970s and 1980s.

INTRODUCTION

"Built like a Mack Truck." What a wonderful statement—one of the best descriptions you can give to a piece of equipment designed to stand up to tough conditions. Since just after the turn of the century, Mack has been turning out trucks of such a quality that the phrase has become part of our language. The first truck placed in the Smithsonian Collection was a Model AC—the venerable "Bulldog."

The Mack brothers made their name as horse drawn wagon builders in Brooklyn, New York in the 1800s. About 1902 they produced their first motor vehicle, and by 1911 Mack produced over 500 trucks a year. The outbreak of World War I proved a boon to their business, which by then was established in its present headquarters in Allentown, Pennsylvania.

The demand for Mack strength and reliability was always strong in the construction and mining industries. By the mid-1930s, the AC and AP were dated vehicles, designed some twenty years before. To remain competitive, Mack designed a new line of off-highway trucks and named them the "Super-Duty" line. The Iconografix title *MACK FC-FCSW-NW 1936-1947 PHOTO ARCHIVE* covers the largest of these trucks, and this book covers the remaining Super-Duty models.

Some 1700 Super-Duty trucks were built between 1937 and 1950. With a few exceptions, all were chain driven and served as dump trucks or concrete mixer units in the construction or mining industries. The FG, FH, FJ, and FK were all basically similar, but with larger tire sizes and beefed-up chassis as GVW increased. We have grouped the photos chronologically within models and given negative numbers where known. Model numbers are taken from the negative, but in some cases are assumed.

We hope that by presenting these fascinating images the reader will be encouraged into further research.

MACK MODEL FG

162 were built between 1938 and 1942. Gross Vehicle Weight (GVW) was 35,000 pounds. The standard engine was the 6-cylinder gasoline 468 cubic inch model CF.

FG. February 1939. (A9183)

FG. February 1939. (A9185)

FG. February 1939. (A9184)

FG. February 1939. (A9181)

FG. February 1939. (A9182)

FG. August 1938. (M3279)

FG. September 1938. (M3369)

FGs. January 1939. (M3791)

FG. January 1939. (M3792)

FG. February 1939. (M3882)

FG. March 1939. (M3945)

FG. March 1939. (M3947)

22

FG. March 1939. (M3949)

FG. January 1940. (M4545)

FG. January 1940. (M4547)

FG fleet. April 1940. (M4749)

FG. April 1940. (M4750)

MACK MODEL FH

265 were built between 1937 and 1941. GVW was 40,000 pounds. The standard engine was the 6-cylinder gasoline 518 cubic inch model EO. The 6-cylinder diesel 519 cubic inch model ED was offered as an option.

FH. May 1938. (A8701)

FH. May 1938. (A8699)

FH. May 1938. (A8700)

FH. May 1938. (A8697)

FH. May 1938. (A8698)

FH. June 1938. (M3072)

34

FH. June 1938. (M3073)

FH. June 1938. (M3137)

FH. June 1938. (M3138)

FH. June 1938. (M3139)

38

FH. July 1938. (M3172)

FH. July 1938. (M3207)

FH. July 1938. (M3208)

FH. January 1939. (M3759)

42

FH fleet. January 1939. (M3758)

FH. May 1939. (M4086)

FH. May 1939. (A9414)

FH. January 1940. (M4542)

FH coal dumper for W. J. Makowski, Mt. Carmel, Pennsylvania, chassis #1223. January 1941. (V2479)

FH and FJ. May 1941. (M5771)

MACK MODEL FJ

322 were built between 1938 and 1943. GVW was 45,000 pounds. The standard engine was the 6-cylinder gasoline 518 cubic inch model EO. The Cummins 6-cylinder diesel 672 cubic inch model HB-6 was offered as an option.

FJ. December 1938. (M3623)

50

FJ. May 1939. (A9407)

FJ. April 1939. (A9357)

FJ. April 1939. (A9356)

FJ. May 1939. (M4095)

FJ. June 1939. (M4198)

FJ. November 1939. (M4494)

FJ. February 1942. (M6376)

FJ chassis #1206 for Automoritz O'Farrill, Mexico. October 1941. (V3503)

58

FJ chassis #1206 for Automoritz O'Farrill, Mexico. October 1941. (V3504)

FJ fleet. March 1942. (M6390)

FJ fleet. March 1942. (M6391)

"'Let George do it!' is the answer to lots of tough hauling problems around Baltimore, Maryland. And the George Transfer and Rigging Company answers the challenge of the heaviest loads with rugged, reliable Macks. The George fleet includes sixteen hard-working Mack trucks. That selection was based on actual experience with Mack performance, dependability and stamina. Those are the qualities that are proving every where that you can't beat a Mack—anywhere!"

"A typical George operation being handled by a Mack FJ Tractor. The load on this low-bed trailer scales a big 42 tons... a Main Reduction Gear manufactured by the Bartlett Hayward Division of The Koppers Company for the Victory ship S.S. Frederick. Big jobs like this are easy for Macks."

(M8869A)

(M8869)

FJ of Andrew Gull Corp., New York. September 1949. (M14127)

FJ of Andrew Gull Corp., New York. September 1949. (M14131)

FJ. New York City subway after World War II. (10001)

FJ. New York City subway after World War II. (10004)

MACK MODEL FK

123 were built between 1938 and 1941. The 4-wheel version was rated at 50,000 pounds GVW, and the 6-wheel version at 60,000 pounds GVW. The 6-wheel version was the only non-chain driven F-series truck. Both came with the 6-cylinder gasoline 611 cubic inch model EP engine as standard, and the Cummins 6-cylinder diesel 672 cubic inch model HB-6 offered as an option.

FK. October 1938. (A8970)

FK. October 1938. (A8969)

70

FK. October 1938. (A8968)

FK. October 1938. (M3489)

FK. November 1938. (M3543)

FKSW 6-wheeler. February 1939. (A9215)

FKSW 6-wheeler. February 1939. (A9217)

FKSW 6-wheeler. February 1939. (A9214)

FKSW 6-wheeler. February 1939. (M3904)

FKSW 6-wheeler. March 1939. (A9270)

FKSW 6-wheel tractor. April 1939. (M3996)

FKSW 6-wheeler. May 1939. (M4089)

FKSW US Navy 6-wheel tractor hauling ship's gun turret. December 1939. (M4515)

FKSW 6-wheeler for measuring drawbar pull/horsepower. July 1940. (M4949)

FKSW 6-wheeler for measuring drawbar pull/horsepower. September 1940. (M5041)

FKSW 6-wheeler for measuring drawbar pull/horsepower. September 1940. (M5042)

FKSW 6-wheeler for measuring drawbar pull/horsepower. September 1940. (M5043)

FK. February 1940. (M4629)

FK. May 1941. (M5733)

FK, Oliver Mining. December 1941. (M6215)

FKSW.

"When the Navy offered one of the big guns from the battleship Texas for display during the Sixth War Loan drive, it could find no takers because of the difficulty of moving the huge 14-inch rifle. To Gerosa Haulage, one of New York's leading heavy haulers, such a job is all in the day's work. Sponsored by the New York Sun, the 70-ton gun was easily moved from the Brooklyn Navy Yard up fashionable Fifth Avenue to its point of display near Radio City by one of Gerosa's FKSW six-wheeled Mack tractors. The complete set-up measured 85 feet 10 inches in length. The gun is one of those that performed so remarkably during the invasion of Normandy, when the Texas gave long-range support to the invading forces. January 1944."

MACK MODEL FN

150 were built between in 1940 and 1941. GVW was 35,000 pounds. The standard engine was the 6-cylinder gasoline 468 cubic inch model CF. The 6-cylinder diesel 519 cubic inch model ED was offered as an option.

FN. June 1940. (V1677)

FN 150-inch wheelbase chassis #1010 with deluxe cab. June 1940. (V1631)

FN 150-inch wheelbase chassis #1010 with deluxe cab. June 1940. (V1630)

FN fleet. January 1941. (M5416)

FN. April 1941. (M5687)

FNT tractor. October 1941. (M6075)

FN tractor. (Z1870)

MACK MODEL FP

365 were built between 1940 and 1942. GVW was 26,000 pounds. The standard engine was the 6-cylinder gasoline 354 cubic inch model EN. The Buda 6-cylinder diesel 468 cubic inch model 6DT was offered as an option.

FP 176-inch wheelbase chassis #1052. February 1941. (V2684)

FP 176-inch wheelbase chassis #1052. February 1941. (V2683)

FP. April 1941. (M5633)

FP fleet. April 1941. (M5628)

FP fleet. April 1941. (M5629)

FP. November 1941. (M6188)

FP. January 1942. (M6317)

AC Bulldog, EG, and FP Mack trucks, Gypsum Lime Co. of Canada. January 1942. (M6322)

FP tractor, Gypsum Lime Co. of Canada. January 1942. (M6321)

FP. April 1942. (M6465)

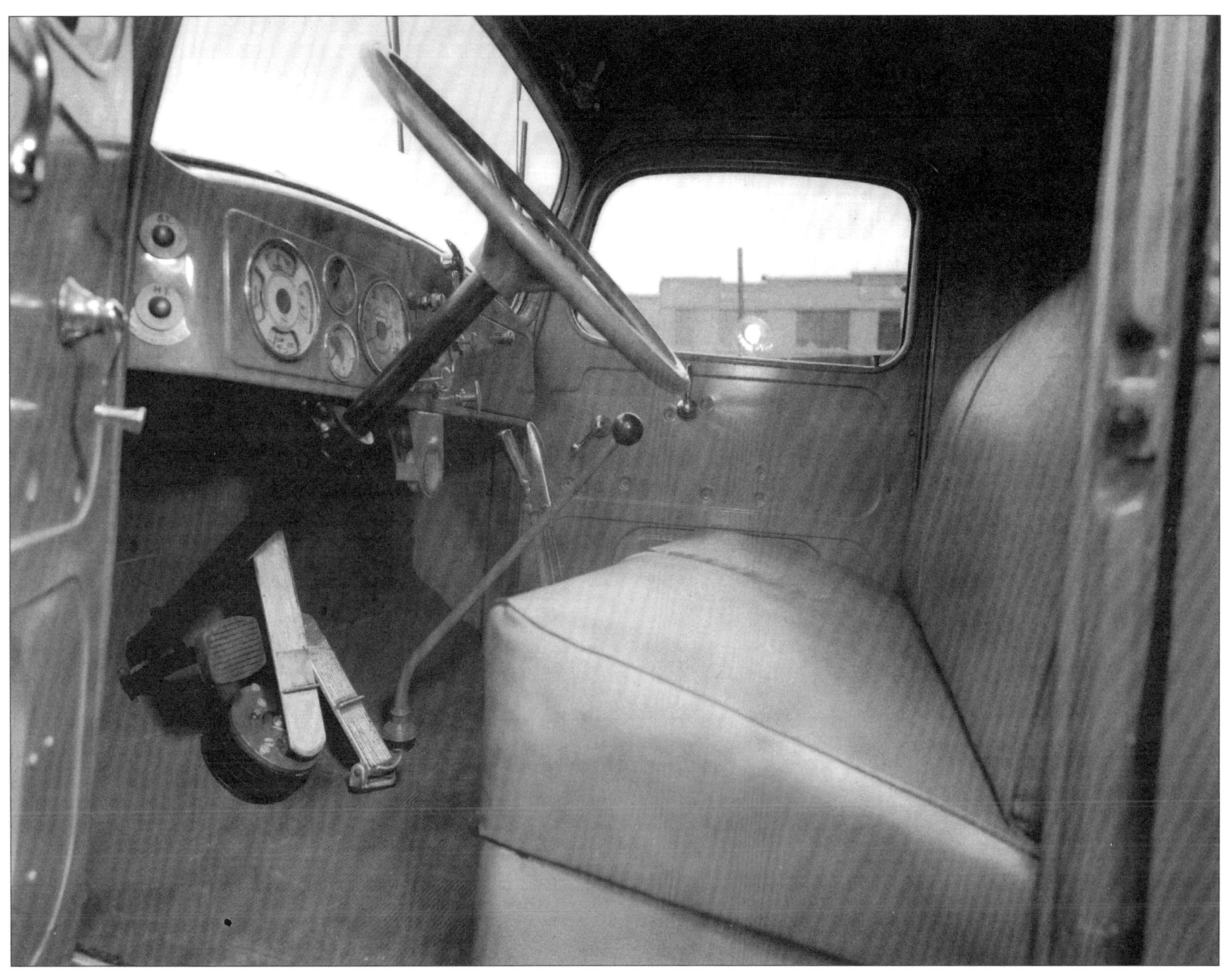

FP. April 1942. (M6464)

MACK MODEL FT

241 were built between 1941 and 1950. GVW was 35,000 pounds. The standard engine was the 6-cylinder gasoline 510 cubic inch model EN510A.

FT1C chassis #1011, Rotundi & Sons. October 1941. (V3471)

FT1C chassis #1290. July 1947. (C1371)

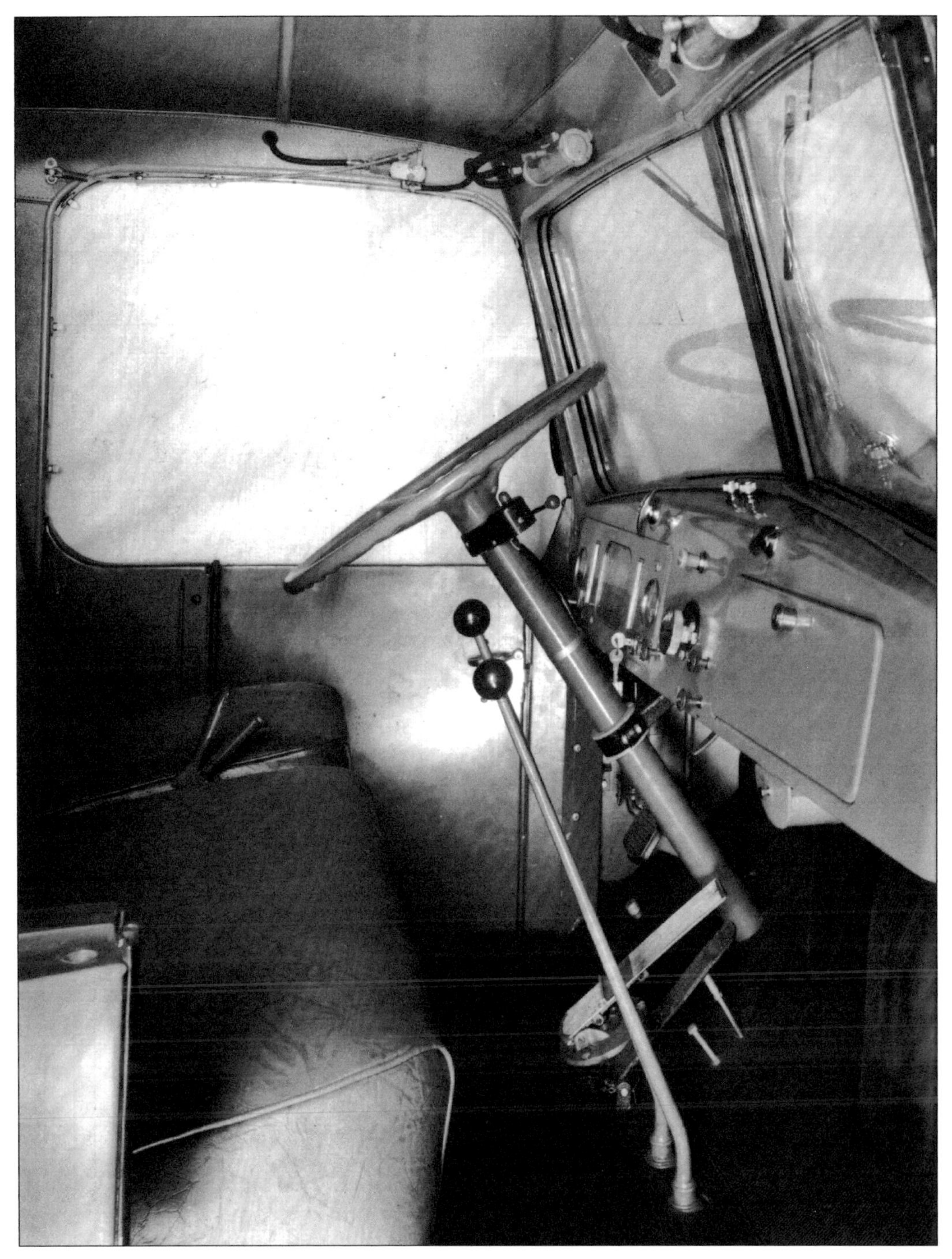

FT1C chassis #1290. July 1947. (C1363)

MACK MODEL FW

This model replaced the FK, and 63 were built between 1941 and 1949. GVW was 50,000. The standard engine was the 6-cylinder gasoline 707 cubic inch model EN707A. The Cummins 6-cylinder diesel 672 cubic inch model HB-6 was offered as an option.

FW1C chassis #1008D for Oliver Mining Co. July 1942. (V4515)

FW1C chassis #1008D for Oliver Mining Co. July 1942. (V4513)

FW1C chassis #1008D for Oliver Mining Co. August 1942. (M6706)

FW. August 1947. (C1493)

120

FW. August 1947. (C1495)

FW. August 1947. (C1496)

122

FW. August 1947. (C1492)

FW, Henry Scheib Co., New York. September 1947. (M11536)

FW, Watson Makowski Coal Stripping Operation. October 1947. (C1724)

BIBLIOGRAPHY

Brownell, Tom, *History of Mack Trucks*, Osceola, Motorbooks International, 1994.

Montville, John B., *Mack*, Newark, Walter Haessner, Inc., 1973.

Montville, John B., *Mack, A Living Legend of the Highway*, Tucson, Aztex Corp., 1979.

Rasmussen, Henry, *Mack, Bulldog of American Highways*, Osceola, Motorbooks International, 1987.

The Iconografix Photo Archive Series includes:

TRACTORS AND CONSTRUCTION EQUIPMENT

CASE TRACTORS 1912-1959 Photo Archive	ISBN 1-882256-32-8
CATERPILLAR MILITARY TRACTORS VOLUME 1 Photo Archive	ISBN 1-882256-16-6
CATERPILLAR MILITARY TRACTORS VOLUME 2 Photo Archive	ISBN 1-882256-17-4
CATERPILLAR SIXTY Photo Archive	ISBN 1-882256-05-0
CATERPILLAR THIRTY Photo Archive	ISBN 1-882256-04-2
FARMALL F-SERIES Photo Archive	ISBN 1-882256-02-6
FARMALL MODEL H Photo Archive	ISBN 1-882256-03-4
FARMALL MODEL M Photo Archive	ISBN 1-882256-15-8
FARMALL REGULAR Photo Archive	ISBN 1-882256-14-X
FORDSON 1917-1928 Photo Archive	ISBN 1-882256-33-6
HART-PARR Photo Archive	ISBN 1-882256-08-5
HOLT TRACTORS Photo Archive	ISBN 1-882256-10-7
JOHN DEERE MODEL A Photo Archive	ISBN 1-882256-12-3
JOHN DEERE MODEL B Photo Archive	ISBN 1-882256-01-8
JOHN DEERE MODEL D Photo Archive	ISBN 1-882256-00-X
JOHN DEERE 30 SERIES Photo Archive	ISBN 1-882256-13-1
MINNEAPOLIS-MOLINE U-SERIES Photo Archive	ISBN 1-882256-07-7
OLIVER TRACTORS Photo Archive	ISBN 1-882256-09-3
RUSSELL GRADERS Photo Archive	ISBN 1-882256-11-5
TWIN CITY TRACTOR Photo Archive	ISBN 1-882256-06-9

TRUCKS

DODGE TRUCKS 1929-1947 Photo Archive	ISBN 1-882256-36-0
DODGE TRUCKS 1948-1961 Photo Archive	ISBN 1-882256-37-9
MACK MODEL AB Photo Archive	ISBN 1-882256-18-2
MACK MODEL B 1953-66 Photo Archive	ISBN 1-882256-19-0
MACK MODEL B 1953-1966 VOLUME 2 Photo Archive	ISBN 1-882256-34-4
MACK EB, EC, ED, EE, EF, EG & DE 1936-1951 Photo Archive	ISBN 1-882256-29-8
MACK EH-EJ-EM-EQ-ER-ES 1936-1950 Photo Archive	ISBN 1-882256-39-5
MACK FC, FCSW & NW1936-1947 Photo Archive	ISBN 1-882256-28-X
MACK FG-FH-FJ-FK-FN-FP-FT-FW 1937-1950 Photo Archive	ISBN 1-882256-35-2
MACK LF-LH-LJ-LM-LT 1940-1956 Photo Archive	ISBN 1-882256-38-7
STUDEBAKER TRUCKS 1928-1940 Photo Archive	ISBN 1-882256-40-9
STUDEBAKER TRUCKS 1941-1964 Photo Archive	ISBN 1-882256-41-7

AUTOMOTIVE

AMERICAN SERVICE STATIONS 1935-1943 Photo Archive	ISBN 1-882256-27-1
IMPERIAL 1955-1963 Photo Archive	ISBN 1-882256-22-0
IMPERIAL 1964-1968 Photo Archive	ISBN 1-882256-23-9
LE MANS 1950: THE BRIGGS CUNNINGHAM CAMPAIGN Photo Archive	ISBN 1-882256-21-2
SEBRING 12-HOUR RACE 1970 Photo Archive	ISBN 1-882256-20-4
STUDEBAKER 1933-1942 Photo Archive	ISBN 1-882256-24-7
STUDEBAKER 1946-1958 Photo Archive	ISBN 1-882256-25-5

The Iconografix Photo Archive Series is available from direct mail specialty book dealers and bookstores throughout the world, or can be ordered from the publisher. For additional information or to add your name to our mailing list contact:

Iconografix
PO Box 609
Osceola, Wisconsin 54020 USA

Telephone: (715) 294-2792
(800) 289-3504 (USA and Canada)
Fax: (715) 294-3414

MORE GREAT BOOKS ABOUT MACK TRUCKS

MACK MODEL AB *Photo Archive*
ISBN 1-882256-18-2
MACK MODEL B 1953-66 *Photo Archive*
ISBN 1-882256-19-0
MACK EB, EC, ED, EE, EF, EG & DE 1936-1951 *Photo Archive*
ISBN 1-882256-29-8
MACK FC, FCSW & NW 1936-1947
Photo Archive ISBN 1-882256-28-X
MACK MODEL B 1953-1966
VOLUME 2 *Photo Archive*
ISBN 1-882256-34-4
MACK EH-EJ-EM-EQ-ER-ES 1936-1950
Photo Archive ISBN 1-882256-39-5
MACK LF-LH-LJ-LM-LT 1940-1956
Photo Archive ISBN 1-882256-38-7

**Available from your favorite bookseller or from Iconografix.
To order by phone:
(800) 289-3504 (US and Canada)
or (715) 294-2792.**

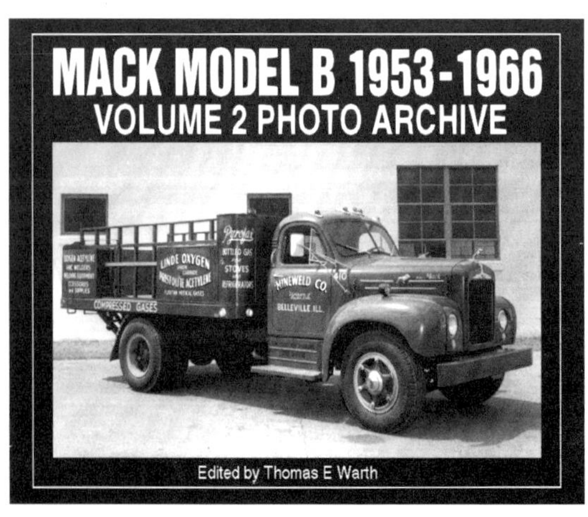

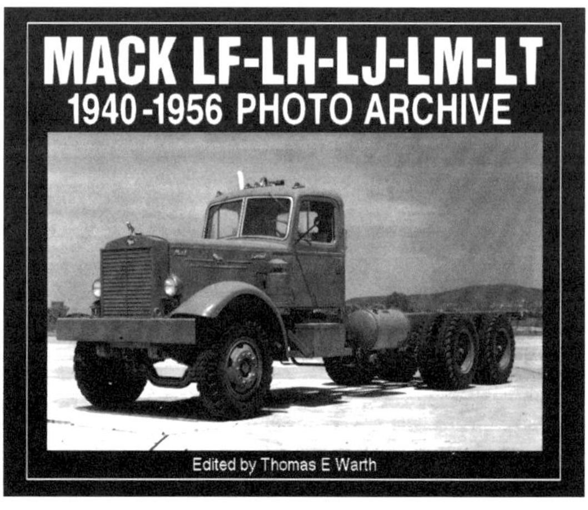